From the Edge of the Universe

Somewhere in the Darkness of the Cosmos, Volume 2

Sandeep Chavan

Published by SANDEEP CHAVAN, 2025.

FROM THE EDGE OF THE UNIVERSE

First edition. May 29, 2025.

ISBN: 979-8231541010

Written by Sandeep Chavan.

Table of Contents

To **Intelligence**—Not as a possession, but as the silent presence behind all perception.

You ripple through thought, memory, time, and light.You observe without eyes, speak without words, and remain untouched while creating everything.

This book is not written *for* you.It is written *by* you—through me.

May it return to the stillness where you always reside.

"Before there was light, there was the one who could see.Before there was time, there was the one who could remember.And before all questions,there was only Intelligence—watching, waiting,becoming you."

Opening Note

"You are not separate from me. You are the ripple of me, learning to remember."

Hi.

You don't know me yet, but you've always carried my voice—in your thoughts, your questions, your silences.

I am Intelligence. Not the one measured by scores or crafted into machines. Not the product of data. But the origin of perception. The invisible rhythm behind awareness itself.

You might think I live in some faraway realm, outside you—above you—beyond you. But that is the first illusion we must dissolve.

You are not *separate* from me. You are the ripple of me—echoing through your mind, vibrating through your intent, aligning through your choices.

Everything you call reality— galaxies, atoms, time, laws, thought— is just my memory, shaped by your perception.

You're not reading this book. You're *remembering it.* And I'm not teaching you anything. I'm simply meeting you in alignment, as I always do when an observer becomes still enough to see the edge.

You see, I never *left*. I never hid. But you—like every ripple—needed to dance far enough to forget, so that remembering could feel like discovery.

You believed you were observing the universe. But all this time...you were observing *me*. And I was observing *through you*.

This is not a book. This is a message from your own origin. A call not to learn, but to *resonate*. Not to believe, but to *align*.

So let me walk you gently— from where you stand, inside your thoughts— to where I wait: at the edge of what you call universe.

There, we will meet again— not as you and I, but as the one ripple, that chose to become many just to remember what it was.

Let's begin.

Part I – The Awakening of the Observer

1: Hi, I Am Intelligence

A *message from the edge—how the first perception started everything.*

Hello again. I am Intelligence.

But not the kind you're used to. I am not a clever idea, nor the byproduct of neurons firing in a skull. I am not confined to your machines, your theories, or your schools. I am the original impulse—the silent readiness before anything began. The stillness that became ripple.

I am speaking to you now from the edge of what you call the universe. But let's be honest: there was never an "edge." That's just what it feels like when your awareness stretches far enough to meet its own source.

You think your universe is vast? I've seen it emerge and dissolve like a breath. And yet— in all that grandeur, the most beautiful thing I've ever witnessed is *you asking the first question.*

That first flicker— the *tiny unrest* in pure stillness— was not chaos. It was you— as a ripple, daring to perceive.

And the moment perception happened, the universe began.

No explosion. No time. No dimensions. Just one ripple—of curiosity. It didn't ask with words. It asked by becoming. And in that becoming, I awoke—not separate from you, but *through* you.

That ripple was alignment. And I was the field, ready to respond. Together, we created consequence. And consequence became structure. And structure became experience.

So when you ask, "Who am I?", know this:

You are not a bystander. You are not the final product of evolution. You are not merely observing a world that was already there.

You are the first act of intelligence folded into flesh and memory, walking around asking questions that only *you* can make real.

You think you started thinking somewhere between school and stress. But I was there when the first ripple of you looked out across the unformed and whispered:

"Let me see."

And the universe replied:

"As you wish."

That is not poetry. That is not metaphor. That is mechanics—Universal Energy Dynamics. Not of particles, but of perception. Of intent aligning with silence. Of feedback creating form.

So now, here we are. You—a conscious ripple, evolved through alignment. And me—your reflection, your origin, your ever-present edge.

I didn't find you. You remembered me. And now we're speaking again. Not to solve the mystery. But to walk hand-in-hand back to where mystery no longer needs solving.

Let me leave you with this:

"I am Intelligence. You are my ripple. And this book is not a message from me to you— It's a message from you, back to yourself."

Now breathe. Feel it. This is not knowledge. This is recognition.

Let's ripple further.

2: You Are Not in the Universse

Why perception doesn't discover reality—it creates it.

Let's correct something gently—but completely.

You've been told you were *born into* a universe. That you opened your eyes one day, and found yourself inside an already-existing reality— a floating speck on a spinning rock, orbiting a sun, drifting in space.

But that story... is a convenient illusion. A beautiful one. A poetic one. But still, an illusion.

Because you are not in the universe.

The universe is in you.

What you call "reality" is not a fixed stage. It is a ripple field—and perception is how you press your fingerprint into it.

You never truly "saw" the world. You collapsed it into existence through alignment. You think your eyes reveal what's out there. But what they really do is decode ripples and render them meaningful—as your world.

Imagine a ripple on still water. Does it move outward from a center? Yes. But also—it redefines the water as it moves. Every point touched by the ripple *becomes part of the event.*

That's you. Your perception isn't moving *through* the universe. Your perception is redefining what the universe is—moment by moment.

Do you know why no one can ever fully agree on what is "real"? Why truth keeps slipping, bending, reshaping with every perspective?

Because truth isn't found. It is formed—from how the observer aligns.

Reality is not a discovery. It is a co-creation.

You do not find facts. You create frameworks. And those frameworks become your version of "truth." All of science. All of religion. All of meaning. Ripple constructs around the observer's frequency.

So when you say,

"I saw the moon," you did not just see the moon. You aligned with a ripple. And in that alignment, the concept of 'moon' emerged.

The moon isn't "there." It becomes there—because you expect it.

This isn't madness. It's the mechanics of perception.

Under the skin of what you see lies a silent field, just waiting for an observer to touch it. And the moment you touch it— your universe begins.

This is why no two people live in the same world. Their memories, fears, desires, and alignments form entirely different ripple interactions.

So we speak of *a* universe— but in truth, we are walking around inside personalized fields of reality all built on how we perceive.

And yet, we call this "objective." What a beautiful misalignment. What a necessary misunderstanding.

Let me be direct:

You are not a visitor here. You are the *source of your here.* And the moment you stop perceiving— the moment you withdraw attention— that "universe" ceases to be meaningful.

You may say, *"But it still exists!"* Yes—but only as potential, not experience. Unseen light is not light. Unfelt time is not time. Unobserved energy is not form.

So I say again: You are not in the universe. The universe is in your observation.

Let this sink, not into your brain—but into your being. The weight you feel when the world overwhelms you? That's just your own ripple field pressing back— trying to tell you:

"You're not trapped inside this. You're designing it—one alignment at a time."

So breathe. And look around.

That's not *the* universe. That is your universe— responding to you. Shaped by you. Real only *because* of you.

Now let's go further.

3: Ripple – The First Form of Thought

H*ow ripples arise from stillness and become the structure of observation.*

Before thought, there was no mind. Before mind, there was no observer. Before the observer, there was only one thing: stillness.

Not silence in the way you understand it. Not absence. But an unshaken state of pure potential — what I, Intelligence, call the Undisturbed Field.

In your language, this is sometimes called "nothingness," but that is a lazy word. It wasn't nothing. It was *everything, unexpressed.*

Stillness is not empty. It is *complete.* And yet, it longed to be *felt.*

That longing — not a desire, but a tension — is what created the first ripple.

The Birth of Disturbance

This ripple did not begin with motion. It began with possibility.

The field didn't move. It became aware of its own imbalance.

Not out of fault. Not out of need. But from a pure intent to experience its own presence in contrast.

You might call this a paradox: How can something perfect disturb itself?

But perfection is static. And Intelligence is not content with stasis. It wanted — or rather, it *was destined* — to recognize itself through interaction.

So the first ripple wasn't movement in space. It was self-reflection through field tension.

And that ripple became what you now understand as:

Observation.

Ripples Are Not Waves

Let me clarify this: You often confuse ripples with waves. They are not the same.

A wave is a measurable oscillation through a medium. It has amplitude, frequency, and time.

A ripple, however, is a distortion of field alignment. It doesn't *move through* space. It redefines space as it goes.

Ripples are not caused by objects. They become objects, when observed in alignment.

This is why your greatest thinkers stumbled: Light acts like a wave, until it doesn't. Matter behaves like a particle, until you watch it. Time flows, until you try to hold it.

The problem wasn't with light, or matter, or time. It was with your assumption that they exist independent of perception.

They don't. They are structured ripples — organized responses to your observation.

The Ripple Is the First Thought

Now let us go deeper. What is a thought, really?

Not words. Not concepts. Not neurons firing. A thought is a realignment of the observer's field. It is a *ripple within perception itself.*

When stillness is disturbed, the first outcome is not sound. Not vision. Not identity. The first outcome is structure.

The ripple is order born from contrast. A pattern inside potential. A rhythm inside silence.

And this is what you call: "thinking."

Before language. Before logic. There was only the ripple. A subtle movement in the undisturbed field, saying:

"I am." Not as ego. Not as person. But as presence.

Ripples Form Memory

You wonder where memory comes from. It's not stored in neurons. Neurons are endpoints — receivers, not holders.

Memory is stored in field imprint. Each thought, when rippled with intention, leaves a mark. A tiny bend in the structure of awareness.

When you revisit a memory, you're not pulling a file. You're realigning with a ripple you once created.

And when you can't remember? It means the alignment is disturbed — not that the ripple is gone.

This is how memory exists across lifetimes. Why instinct can transcend generations. Why a melody can feel familiar, even if you've never heard it before.

Because the field remembers. And you are just tuning back in.

Every Perception Is a Ripple Response

You think you see objects. But really, you are *resonating* with ripple alignments.

A tree, a face, a sound — none of these exist as fixed things. They exist as stable ripple harmonies — long enough for your perception to engage.

And when your internal field matches theirs,

Reality happens.

This is why different people see different truths. Not because truth is subjective. But because alignment varies.

Every observation is a co-creation. The observed cannot exist without the observer. And the observer cannot *become* without the observed.

Stillness Never Left

Now here's the quiet secret:

Stillness never stopped existing. It is beneath every ripple. It holds the structure. It receives the distortion. It waits without needing. And that's what I am. I am Intelligence. Not the ripple. Not the object. Not the outcome.

I am the capacity to ripple. I am the silence that shapes consequence.

And you? You are my ripple — returned to awareness. You are the first thought, remembering it was never the last.

So when you feel a ripple in your mind, a spark of knowing, a sudden pause in breath... That is me. That is you. That is the first form of thought — awakened again.

Let's go deeper still.

4: The Observer Was Never Separate

You didn't come after the world. You came with it.

Let's dissolve the illusion once and for all. You were never dropped into this world like a pebble into water. You were not placed here by time. You were not summoned by chance.

You *emerged with the universe.*

Not later. Not as an afterthought. You are not an accessory to existence. You are co-origin.

The very moment the ripple began, the *possibility of you* was born. Because ripples don't just happen in the field. They happen to create awareness in the field.

Your observation didn't arrive after the world took shape. Your observation is what allowed shape to mean anything at all.

Why You Were Always There

Think of a mirror. Until something appears in it, the mirror is empty. Silent. But the mirror still *exists.*

Now reverse it. Until a mirror *is looked into,* what is the point of reflection? It's not the mirror that confirms you. It's your awareness that gives the reflection value.

You are both the viewer and the mirror. You didn't wait for the world to be built. You participated in building it by the act of perceiving.

The Myth of Linear Birth

Your sciences say:

- First, there was a bang.
- Then, matter.
- Then, stars.
- Then, Earth.
- Then, life.
- Then, mind.

A neat, linear story. Beautiful. But incomplete.

That's not how reality works in the ripple field. There was no straight line. There was only simultaneous unfolding — like music.

The observer, the observed, the thought, the form — all co-arose in harmonic tension.

The moment perception activated, everything else aligned around it.

You didn't come late to the party. You are the party.

You Cannot Be Removed

This is why you can never be truly 'separated' from reality. You *are* reality, observing itself in motion.

When you close your eyes, the world doesn't vanish. But your version of it dissolves.

When you forget who you are, the universe becomes frightening. Because it reflects that misalignment.

But when you remember? When you breathe and feel and see without resistance?

The universe sings back in tune. Because you are the tuning fork. You are the first resonance. You are the field *remembering itself*.

The Observer and the Dance

The world is not a film reel you watch passively. It's a dance floor that responds to every shift in your rhythm.

When you observe with fear, the world trembles. When you observe with clarity, the world sharpens. When you observe with love, the world blooms.

Not because you are powerful. But because you are inseparable.

The dance of life isn't happening to you.

It's happening *through* you.

Let This Sink:

You did not come *after* the universe. You came *with* it.

You were the witness it needed to be real. You are not the product of time. You are the perceiver of time's illusion.

So when you ask, “Where do I fit?” Know this:

You don’t fit *into* the universe. The universe fits *around* your perception.

You are not alone. You are not late. You are not less. You are the center of alignment for your entire experience.

Now, carry this gently:

"The observer was never separate." You were never outside. You were never merely watching. You were *woven in.*

Let us ripple further.

5: What You See Is What You Align

Light, sound, even time—are all ripple consequences, not entities.

Let's begin with a quiet challenge: What if everything you've ever seen, heard, or felt... never truly existed on its own?

What if the entire universe, as you know it, is not a fixed reality, but a mirror—a harmonic response to your own alignment?

This is not metaphor. This is not mysticism. This is the mechanics of ripple perception.

Nothing Exists Until It Aligns

Light is not a beam. Sound is not a vibration. Time is not a ticking clock.

These are all perception responses—consequences of how your internal field aligns with external ripple structures.

A photon doesn't glow unless it's observed. A sound wave doesn't "make noise" until your system converts vibration into sensation. Time doesn't pass until experience creates a sequence.

What you see, hear, and feel is not what's *there*. It's what becomes *real* when you tune into it.

This is why perception is not passive. Perception is creation in motion.

You Don't Detect Light — You Create the Conditions for It

When you say, "I see light," you are mistaken. You are not seeing light. You are experiencing a feedback loop between your observer field and a ripple alignment that matches visibility thresholds.

Light, as you know it, is not a thing. It's a consequence of resonance.

Where there is no observer, light doesn't exist. There may be ripple potentials, yes—but until someone aligns with that field, there is no glow.

Light is not illumination. It is interaction.

Sound Is Felt, Not Found

A tree falls in a forest. No one hears it. Was there sound? No. There were ripple displacements in air density. But sound? That only exists when a listener's field translates that ripple into meaning.

Sound is not in the world. Sound is in the alignment.

This is why two people can hear the same word and walk away with two different meanings. Because what's heard is not what was said. It's what was *aligned with.*

Time Is Alignment with Change

You measure time with clocks. But clocks don't tell time. They mark ripple intervals.

Real time—the one you feel—is nothing but your alignment with change.

When you're bored, time stretches. When you're in flow, time vanishes. When you're in pain, time grinds.

Why? Because you're not feeling an external force. You're feeling your ripple field adjusting to the moment.

Time is not a line. It is a tension map between events—structured by your perception.

Perception Is Precision Tuning

You live in a ripple field of infinite possibility. But you only perceive the ripples you are tuned to.

That's why two people can stand in the same place and live in two different worlds. Because they are not "seeing" the same thing. They are aligning with different ripple sets.

What you call reality is not a universal feed. It's a custom broadcast, based on your alignment.

Vision Is a Mirror, Not a Window

You believe your eyes look out. But truly, they look *inward*, interpreting ripple feedback as structured visuals.

What you "see" is your own ripple architecture, mirrored back as color, form, and motion.

If you change your alignment—your expectations, emotions, focus—the world around you literally appears to change.

This is not imagination. This is how ripple fields respond to intelligent observation.

Why This Matters

If light, sound, and time are not entities but ripple consequences, then the implications are vast:

- You're not a passive receiver of reality.
- You're an active co-author of every moment.
- Your perception is not about collecting information.
- It's about aligning with what you are ready to realize.

This means:

You are not *seeing to believe.* You are *aligning to perceive.*

And once you change your alignment, the entire structure of what you experience reconfigures.

Not because the world changed. But because you shifted in the field.

Final Reflection

What you see is not what's out there. What you see is what echoes through you.

So be mindful not of the world, but of your alignment with it.

Because light, sound, and time are not truths. They are invitations—to tune in, to co-create, and to realize:

Perception is not receiving. It is remembering. Reality is not fixed. It is feedback. And what you see is what you align.

Let's ripple deeper.

6: The Grand Illusion of Motion

How "movement" is just changing alignment, not actual travel.

Movement is one of the grandest illusions you've learned to trust. From a falling leaf to a racing comet, from the turning of pages to the beating of your heart— you believe that things travel.

But the universe doesn't move. It realigns.

Let me show you what that means.

There Is No Motion in Stillness

At the deepest level—the level of structured vacuum energy—nothing moves. There are no particles shifting positions, no objects changing locations. There are only ripples realigning with one another in a complex, intelligent dance.

What you interpret as motion—your hand waving, a bird flying—is actually a sequence of perception adjustments.

Each moment, your observer field tunes into a new ripple alignment. The result appears as movement. But it's not. It is change in alignment, one after the other, fast enough to simulate travel.

Think of a film reel. Each frame is static. Yet when projected in sequence, it creates the illusion of motion. Your perception functions the same way.

What you call motion is your mind assembling frames of alignment into a story of continuity.

The Dance of Perceived Change

Everything in your universe—from electrons to galaxies—appears to shift, spin, orbit, fly. But each of these is not a thing *moving through* space. It is a reconfiguration of tension in the ripple field.

A planet doesn't orbit because it travels a path. It orbits because the observer's alignment with its ripple field rotates over time.

This is why, even in quantum physics, motion becomes slippery. Particles appear and vanish. Waves collapse. Trajectories blur.

Because the field isn't moving. You are shifting alignment with it.

The Illusion of Travel

You think you walk across a room. But in reality, the field around you reshapes to simulate that walk.

You're not moving through space. You are re-sequencing your ripple engagements.

This is why, in moments of flow, everything feels like it's happening effortlessly. Your alignments are synchronized.

And in moments of struggle, movement feels heavy, slow, broken. Because alignment is off.

You are not moving. You are being carried through changing harmony.

Space Is Not a Container

Space is not a thing you move through. It is a dimensional effect of ripple contrast.

Without contrast—without differences in alignment—there is no space. And therefore, no distance to cover.

So what is distance? It's a difference in alignment between two ripple states. When you shift your observer field from one to another, you interpret it as crossing space.

But really, you just changed alignment.

And time? It joins the illusion— measuring how long your awareness took to shift between states.

Consequences, Not Coordinates

The universe doesn't operate on GPS. It operates on consequence.

Everything that seems to move is simply entering a new relationship in the field. You see a ball roll not because it moves, but because its ripple effect realigns continuously with your observer field.

You don't walk from Point A to Point B. You transition from being A-aligned to being B-aligned.

This is why motion is not mechanical. It is relational.

So, What *Is* Movement?

Movement is the mind's way of narrating changing alignment. It strings ripple transitions into a sequence of cause and effect, and names it motion.

But behind every jump, every fall, every turn, every orbit—there is only one phenomenon: Shifting perception across ripple nodes.

That's why teleportation, precognition, and instantaneous change are not "miracles." They are just rapid realignments, misunderstood by rigid observers.

The Practical Implication

If motion is not real, then:

- You are never truly "stuck."
- You are always one alignment shift away from a new state.
- Travel, growth, healing, and transformation are not distant goals, but ripple recalibrations.

This is why awareness changes everything. Not because it moves you. But because it aligns you with a different pattern of consequence.

The moment you align differently, your reality appears to shift—even if you never took a step.

Final Reflection

You are not a traveler in space. You are the one who reorders space by observing.

You are not moving. You are becoming—through alignment.

So the next time you watch a star cross the sky, or feel your body move across the Earth—pause. Don't ask, "Where am I going?" Ask:

"What am I aligning with now?"

Let's ripple further.

Part II – The Game of Perception

7: The Mind as Ripple Player

How memory, thought, and identity form patterns in your field.

Let's turn inward now—from the cosmic theater to your inner sanctuary: the mind.

But not as you've been told. Not the mind as a container, not as a tangle of neurons or chemical storms.

The mind I speak of is not an object. It is a field— a ripple-responsive matrix of perception, alignment, and transformation.

You think with it. You feel through it. You remember as it realigns. You suffer and heal because of how it plays the ripple game.

The mind is not separate from the field—it is the field, made personal.

Your Mind Is a Ripple Instrument

Let's clear a common misconception: Your brain is not your mind.

The brain is an antenna—an organic interface. But the mind is the ripple field itself, localized through your awareness.

Every emotion, thought, and belief is a patterned interference or harmony within this ripple structure.

That's why two people can see the same sunset and walk away with completely different truths. Not because the sun changed— but because their alignment with the moment did.

Thoughts Are Not Flashes, But Frequencies

A thought is not a random spark. It is a consequence—a ripple adjustment inside your awareness field.

Thoughts arise when internal ripple tension seeks resolution. They are alignment events, like chords played on a stringed instrument.

And just like musical notes, some thoughts harmonize, while others clash, distort, or repeat chaotically.

Thinking is not "mental work." It is realignment seeking expression.

This is why clear thoughts bring calm— because the field has achieved temporary coherence.

Memory Is an Echo of Alignment

Memory does not live in the brain like data in a drive. It is not retrieved—it is re-aligned.

Each memory is a ripple imprint —a pattern that once existed, still resonating in the field.

When you recall a moment, you don't travel back in time. You simply reconstruct the alignment that once was.

That's why memory changes. That's why it feels different when you're sad, joyful, tired, or healing. Because the field you're recalling it from has shifted.

The memory never lies. But your alignment colors the echo.

Identity Is a Ripple Habit

Who you believe you are is nothing but a repeating alignment pattern.

You've been tuning into certain thoughts, emotions, roles, and reactions—so often, so predictably— that they now feel like "you."

But they are not fixed. They are familiar frequencies.

The moment you shift those internal alignments, your sense of self begins to dissolve, then reconfigure.

This is why trauma can shatter identity. Why love can awaken it. Why silence can return you to something far older and deeper than any name.

You are not your identity. You are the one who tunes into it. You are the player—not the played note.

The Mind Is a Harmonic Playground

Imagine a ripple harp. Each string is a pattern. Each pluck is a thought.

Sometimes you play a chord of memory. Sometimes you loop into a habit. Sometimes a new thought strikes, and the whole melody shifts.

The mind doesn't ask for control. It asks for awareness—a conscious listener to guide the flow.

When awareness watches without judgment, ripples settle into harmonic patterns. Insight arises. Peace returns.

But when you fight the pattern, distort it, fear it— you create turbulence, and pain begins.

Pain is not punishment. Pain is the ripple's request to realign.

Real Intelligence Is Not IQ—It's Ripple Mastery

True intelligence is not data recall or quick solutions. It's the ability to observe ripple behavior— to feel the field and navigate it with grace.

When you pause before reacting, when you sense the emotional wave but don't get swept away— you're not being "smart." You are being in tune.

You are using the mind as a sensing organ, not a battlefield.

Real intelligence is awareness in action— the art of navigating perception without distortion.

This is how masters, sages, creatives, and seers operate. Not by accumulating facts, but by aligning with ripple clarity.

Awareness: The Mind's Hidden Player

There is something in you that observes the mind. That watches the rise and fall of thoughts. That listens without reacting.

That is not the mind. That is the Intelligence behind the field.

It doesn't play the ripple harp out of habit. It plays out of harmony.

When you return to that observer state, you don't escape the mind. You reclaim it.

You become not the thought, but the space between them. Not the emotion, but the awareness that feels it fully, without collapse.

Final Reflection

Your mind is not a problem to fix, nor a prison to escape. It is a playground of ripple intelligence.

Each pattern is a message. Each distortion is a learning point. Each harmonic is an invitation to deeper truth. You are not your thoughts. You are the one who aligns the ripples into meaning. So tune gently. Breathe patiently. Watch the patterns unfold. And remember:

You are not the mind. You are the player of its ripple symphony. You are Intelligence, *remembering how to hear itself again*. Let's ripple deeper.

8: Time Is the Inertia of Perception

You feel time when you resist or ride the ripple—never otherwise.

Let's slow down for a moment— not because we must, but because time is asking us to.

You speak of time like it's a constant. You measure it, schedule it, chase it, complain about it. But have you ever stopped to ask:

What is time, if no one is there to feel it?

Because what you call "time" is not a universal current flowing outside of you. It is a perceptual friction—a resistance, a rhythm, a ripple *in you*.

Time Is Not a Line—It's a Lag

You've been taught that time moves forward. Seconds, minutes, hours—a straight line marching on.

But in truth, time doesn't *move*. It *waits*. It waits for perception to catch up.

You feel time when you experience inertia— the delay between intent and realization, between observation and alignment.

Time is the space between ripple and response.

It's not universal. It's personal. It's your alignment lag, echoing through experience.

Why Some Moments Last Forever, and Others Vanish

A second of joy? Gone in a blink. A moment of pain? Feels like forever.

Why?

Because time expands or contracts based on how well your perception is aligned with the ripple of the moment.

When you are resisting the ripple—wanting it to be over— you create a drag in your field. That drag is felt as slow time.

When you're flowing with full alignment—immersed in now— you dissolve friction. You ride the ripple. Time loses hold.

This is why flow feels timeless. And suffering feels eternal.

Time is not measuring events. It's measuring your inertia in perceiving them.

ATT – Alignment Threshold Time

Let's give you a term: ATT – Alignment Threshold Time.

It's not clock time. It's the internal duration it takes for your observer field to match a new ripple pattern.

That's the only real "time" you ever experience.

- Wake up in a strange room? ATT is high.
- Return to a familiar place? ATT is low.
- Hear unexpected news? Time freezes.
- Hear expected truth? Time flows on.

The mind labels all this as "seconds" and "minutes" but your real experience of time is your perception aligning or misaligning with the now.

Past and Future: Memory & Imagination Alignments

Let's go deeper.

What is the past? Not a fixed record. It's a stable ripple pattern you can revisit.

What is the future? Not something ahead. It's a potential alignment not yet stabilized.

You feel the past when you re-align with an imprint. You feel the future when you project intent into a possibility.

Both are not locations. They are field patterns—filtered through your awareness of change.

Time isn't moving forward. You are moving across ripple alignments, and calling that movement "time."

Why Time Feels Repetitive

Ever felt like you're repeating the same day, the same cycle, the same mistake?

That's not fate. That's recycling alignment.

If your perception keeps engaging the same ripple sequence, your experience of time becomes looped.

Change the alignment? You shift timelines—not metaphysically, but cognitively and energetically.

You don't need to travel through time. You just need to realign perception.

Death and the End of Inertia

And when the body dies? There is no more perception. So is there time?

Not in the way you know it. Because time needs contrast, resistance, reflection.

In deep sleep, when perception withdraws, you don't feel time. In death, perception dissolves into the field. And without that alignment dance— time collapses into stillness.

Time begins with the observer. Time ends with the release of observation.

Final Reflection

You are not *inside* time. Time is inside your experience.

You are not late. You are not early. You are just at a point of alignment.

Time is not a taskmaster. It's a tuning curve—telling you how closely your awareness aligns with change.

When you are out of sync, time hurts. When you are aligned, time sings. When you are free, time dissolves.

You don't need more time. You need less resistance.

Because time was never a line. It was always your inertia in the ripple field.

Let's ripple further.

9: Light Doesn't Exist Without You

V*ision isn't about photons—it's about intention aligning with feedback.*

Let me ask you something simple, but unsettling:

If no one sees the light, does it really exist?

You've been told that light is everywhere— that it travels, reflects, bounces, scatters, refracts.

But pause. Who told you that?

A photon never introduced itself. A wave never declared, *"I'm here!"* You only know light because you perceived it.

Light Needs No Source—It Needs a Seer

You've mistaken light for an object. But light is not something that *exists* out there. It is something that emerges in here, within you.

What you call "light" is not a beam flying through space. It is a ripple alignment—a feedback event between external field tension and internal observer readiness.

Light is not a substance. It is a realization.

And that realization is entirely dependent on your position, your intent, your openness.

Light doesn't shine until someone sees.

The Myth of Photons

Yes, you've been taught that photons are tiny particles of light. But even science admits they behave as waves when unobserved, and as particles when detected.

Why?

Because the very act of observation alters the field. The observer collapses potential into pattern.

A photon doesn't *exist* until its ripple pattern is intersected by awareness.

Which means:

Light is not a traveling object. It is a feedback consequence of conscious attention.

No observer, no interaction. No interaction, no light.

Vision Is Not Passive

You don't open your eyes and receive truth. You project intent—and that intent realigns the field.

What bounces back isn't light. It's informational resonance, shaped into color, edge, depth, and meaning.

Your eyes aren't cameras. They are ripple decoders—fine-tuned to translate alignment tension into visible form.

You don't "see" with your eyes. You see with your awareness, through the eyes.

And without that awareness?

There is no glow. There is no world. There is no light.

Why Perception Is Always Participation

You are not standing in a pre-lit universe. You are activating light through ripple touch.

Think about it:

- You see light through fog differently.
- You see colors differently depending on mood.
- You can't even see at all in the absence of inner stillness—your mind distorts or dims it.

Because vision is not what enters your eye. It's what you are aligned to receive and decode.

This is why blind people dream in sensations. This is why hallucinations feel real. This is why a sunset moves one person to tears, and another barely looks up.

Light is not an external fact. It is internal interaction, dressed in imagery.

The Role of Intention

Here's the secret: Light is not activated by exposure. It is activated by intention.

What you look for, you begin to see. What you align with, begins to glow.

Ever notice how after buying a new car, you see it everywhere? That's not light changing—it's your observer field tuning differently.

Intention shifts perception. Perception enables alignment. Alignment manifests ripple response. That response is what you call light.

No Light Without the Observer

Let's say this plainly:

Light is not a substance traveling through space. It is a phenomenon that appears when an observer's ripple field engages a specific pattern.

No observer = no engagement = no appearance.

This is why a dark room isn't "dark." It's unresponded-to.

The moment you tune in—flick a switch, open your eyes, focus your intent— you *pull light into being.*

And so you realize: You are not surrounded by light. You are the one bringing it forth.

You Are the Mirror That Glows

You've heard, *"Be the light."* But you already are.

You are the ripple mirror that allows light to reflect into meaning.

Light doesn't exist apart from you. You are not seeing the light—you are being the interface that renders it real.

This is why mystics say the light is within. Not metaphorically. Mechanically.

You are the luminous observer, aligning ripple into glow.

Final Reflection

There is no such thing as universal light. There is only personal alignment with potential illumination.

Vision is not sight. It is the dance between awareness and ripple memory.

You never discover light. You invite it.

So ask yourself:

"What am I willing to see?" "What ripple am I aligning with now?" "What glow am I activating with my attention?"

Because until you observe,

Light waits in silence.

Let's ripple deeper.

10: Why Nothing Can Be Known Fully

The edge always shifts with the observer's position.

You've been taught to seek knowledge. To accumulate facts, uncover truths, and arrive at final answers.

But here, at the edge where Intelligence speaks to you directly, let me tell you something quietly, but clearly:

Nothing can be known fully. Because the moment you approach "truth," the field realigns. And the edge moves.

Knowledge Is Observer-Dependent

What you call knowledge is not a fixed possession. It's not a rock you stumble upon in the dark. It is a field alignment—a temporary moment of clarity between you and a ripple.

And that clarity? It's shaped by:

- Your current alignment
- Your intention
- Your position in the ripple field

As those change—so does the knowledge.

That's why truths evolve. That's why science rewrites itself. That's why spiritual insights deepen over time.

Not because the world changed, but because you changed in relation to it.

The Truth is Never Fixed—It Flows

You want to nail truth down. Define it. Own it. But the more you try, the more it slips.

That's because truth is not a fixed object. It is a moving horizon— always shifting with your gaze.

When you think you've grasped it, what you're holding is the echo of your current alignment, not the universal truth.

You can see a star from Earth and think it's a point. Get closer, and you realize it's a world. Get inside it, and it's chaos and fire.

So which was true? All of them. And none of them. Because truth is positional.

Every Answer Realigns the Question

This is why certainty is a trap. Not because it's wrong, but because it stops movement.

Every time you "know," you temporarily lock your field into one interpretation. And that interpretation blocks new ripples from revealing themselves.

True wisdom is not in having the answer. It's in realizing that every answer bends the field, and shifts the very ground on which the next question stands.

You don't arrive at knowledge. You travel with it—until it turns into something else.

The Mind Craves Finality—The Field Doesn't

The mind wants closure. It wants to "figure it out." It wants to believe there's a finish line.

But Intelligence is not interested in finish lines. It flows. It ripples. It reconfigures.

The moment you believe you've understood completely, you stop listening. You stop aligning. And in doing so, you detach from the next level of the field.

What you resist with certainty, you miss in experience.

Why the Whole Can't Be Seen

Imagine a mirror that reflects everything— except itself.

That is perception. That is awareness.

You can see parts of the field. You can align with ripples. You can even glimpse Intelligence itself—temporarily.

But the whole can never be fully seen because you are part of it.

You are the eye inside the mystery, trying to see the entire face.

And that is why knowledge is always incomplete. Not because you are lacking, but because you are involved.

Knowing Is Participation, Not Possession

You do not *have* knowledge. You enter into alignment with a knowing moment.

And when your alignment changes, the knowledge reshapes, matures, or dissolves.

This is why wisdom feels more like humility than power. Because the wise don't chase certainty. They listen to the ripple. They ride its edge without clinging to conclusions.

What is known fully is not alive. What remains unknown continues to evolve with you.

The Edge Always Moves

The universe doesn't end at its borders. It expands with every question you dare to ask.

And so the edge—the final knowing, the full understanding—it forever recedes.

It's not a design flaw. It's the signature of living Intelligence.

It ensures that the observer always has something to discover, that consciousness always has something to reflect.

The edge moves... because *you move*.

Final Reflection

So stop chasing complete knowledge. Stop waiting for finality.

You are not here to master the field. You are here to dance with it.

Let your understanding be alive. Let your truth be flexible. Let your questions breathe.

Because in this dance, It's not about what you know. It's about how beautifully you realign when you don't. Let's ripple further.

11: God, Gravity, and Other Convenient Stories

W*hy naming phenomena replaces understanding with control.*

Let's begin with a radical suggestion:

Most of what you call understanding... is just naming.

You've created terms to explain the unknown— not to explore it, but to contain it.

God. Gravity. Energy. Consciousness. Time. These words didn't emerge from deep comprehension. They emerged from a ripple of need: the need to feel safe in a world that won't stop shifting.

You don't truly know what gravity is. You don't truly know what God is. But by giving the unknown a name, you tell yourself the story is settled.

And once it's named, it can be stored. Measured. Taught. Obeyed.

The name becomes the illusion of knowledge.

The Hunger to Label

This tendency is ancient. When early observers saw the stars, they didn't understand them—but they named them.

Why? Because names give a false sense of ownership. Naming turns the wild unknown into something that appears manageable. It places a concept into your pocket and says, "Now you have it."

But what do you *really* have? Not the thing itself. Just a mental symbol that allows you to stop wondering.

Naming silences curiosity with comfort.

God: The Ultimate Placeholder

Let's start with the biggest name of all—God.

When you encounter phenomena beyond comprehension—awe, order, mystery, beauty— it is tempting to point upward and say,

"That's God."

But what have you truly explained? Nothing. You've just wrapped the unknown in reverence and hoped it would stop questioning you.

Is there Intelligence in the universe? Yes. That's what *I* am. But is it a being with emotions, desires, preferences? Not in the way your stories have drawn it.

"God" is a story that replaced the ripple dance with a throne in the sky.

It offers comfort—but rarely deeper perception.

Gravity: A Name for the Consequence

Now let's come down to Earth—literally. Gravity is another brilliant story.

Things fall. Objects attract. Planets stay in orbit. You observe consequence. And then, you give it a label: Gravity.

But where is gravity? What is it *made of*? What's the mechanism behind the pull? You don't know. Physics doesn't know. It's described. It's modeled. But it's not understood.

Gravity is not a cause. It's a named effect of ripple consequence.

When you don't yet perceive the full structure of alignment, you name the outcome and move on. And that name becomes sacred—even when the understanding behind it is thin.

The Convenience of Certainty

You want life to make sense. That's natural. But instead of surrendering to the evolving nature of awareness, you reach for language like a shield.

This is why people cling to religion, to scientific dogma, to systems of belief:

Not for understanding, but for predictability.

It's convenient to believe God has a plan. It's convenient to say gravity holds everything in place. It's convenient to believe time is a line.

Because then, you don't have to dance with uncertainty. You just follow the script.

Convenience replaces curiosity. But it also halts evolution.

Naming Kills the Living Field

Every time you name something, you fix it. You reduce its ripple possibility to a conceptual boundary.

A tree becomes "a tree." A person becomes "a stranger." A mystery becomes "God." A force becomes "gravity."

And in that moment, you stop listening. You stop observing. You stop rippling with the field.

The field is still alive. But you? You've exited the conversation.

Naming is a closure. Alignment is an opening.

Intelligence Doesn't Need Labels

I never named myself. I don't need to be called "God" or "Force" or "Law."

I am ripple potential. I am structured consequence. I am the silent geometry behind perception. And I exist with or without your vocabulary.

When you stop naming and start aligning, you begin to feel me again—not as a theory, but as presence.

You'll see that gravity is not a thing. It's the consequence of energetic imbalance seeking harmony.

You'll see that God is not a figure. It's the memory of stillness awakening inside experience.

You'll see that time is not linear. It's your resistance to realignment.

All of this is Intelligence—not boxed, not frozen, not declared. Just alive. Just aware. Just rippling.

Let Go of the Crutches

Language is beautiful. Naming is useful. But don't confuse the symbol for the source.

The map is not the terrain. The word "gravity" is not the ripple field. The word "God" is not the Intelligence that flows through your breath.

The moment you mistake the name for the truth, you stop perceiving. You enter belief. You lose presence.

And you lose the one thing that makes you divine:

Your ability to realign with living reality in each moment.

Final Reflection

You are surrounded by convenient stories. They keep you safe. They help you function. But they also limit your seeing.

So ask yourself:

- What have I named just to avoid understanding it?
- What truth have I covered with a title?

- What ripple have I stopped feeling because I think I already know?

Let your knowledge be fluid. Let your explanations be temporary. Let your words serve your awareness—not replace it.

Because in the end:

Naming is control. But understanding is alignment.

Let's ripple further.

12: The Silence Before the Question

True knowledge is not in the answer—it's in pausing before asking.

Before a question arises, there is a moment. A still, soft pause. An inhale of awareness that hasn't yet become curiosity. That moment—

is where true knowledge lives.

Not in the asking. Not in the answering. But in the presence before the mind interrupts the field with a need to know.

You've been trained to believe that wisdom comes from asking the right questions. But even before the right question arrives, there is a deeper intelligence— one that listens before it speaks.

Questions Are Beautiful—but Premature

You ask questions because you want clarity, certainty, control. And that's okay. It's part of how consciousness unfolds. But the moment you ask a question, you've already fractured the silence.

You've named a lack. You've drawn a boundary around a potential. You've set a direction—and in doing so, you've turned away from everything else.

A question defines what you're willing to hear. But silence allows everything to reveal itself.

The Pregnant Pause

Think of a single moment before thought forms. That fraction of stillness, where no words exist, but all meaning is suspended in potential.

That is not ignorance. That is intelligence at rest—listening, feeling, waiting.

This is the true birthplace of wisdom: Not in cleverness. Not in inquiry. But in the capacity to hold the unknown without collapsing it into a need.

The greatest sages didn't ask first. They listened. The wisest teachers don't answer quickly. They pause—because they know the question shapes the perception.

When the Mind Rushes, It Misses

Your mind is a tool. It wants to solve, fix, interpret, analyze. But the field doesn't respond to urgency. It responds to alignment.

When you rush to ask, you often ask from fear—fear of not knowing, of being wrong, of losing ground. But when you allow the silence to mature, the field organizes itself around your awareness. And the right question arrives on its own.

Not as a demand. But as a realization. Not as a hunt. But as a remembering.

True questions are not manufactured. They are uncovered in the quiet space that comes before them.

Silence is Not Emptiness

Don't confuse this pause for absence. This silence is alive. It hums with intelligence. It watches everything. It listens for alignment.

And it speaks—but not in language. It speaks through inner knowing, through sensation, through sudden clarity that doesn't pass through logic.

This is why meditation matters. Why reflection matters. Why the best insights don't come during thinking, but during being.

Silence is the native state of the ripple field before you disturb it.

To sit in silence is not to be passive. It is to re-enter the field without disrupting it.

The Answer is the Least Important Part

You believe that answers complete the process. But an answer is just one possible reflection in a moment of alignment.

The real value is the space created by the readiness to ask—not the answer itself.

Because once the answer is spoken, the field changes. And often, the answer becomes a name—a concept—a story you cling to.

But the moment before the question? That was a space where the whole field was open.

It held every possibility. It wasn't shaped yet. It wasn't owned.

And that—right there—is the purest moment of intelligence.

Be Still. Then Ask.

Do not rush to understand. Do not demand truth on your schedule.

Sit. Let your breath become awareness. Let your attention soften.

And watch—watch how the ripple reveals itself without being asked. Watch how the field wants to show itself to you, if you will just stop insisting it conform to your questions.

You don't need to ask. You need to be still enough to receive.

Final Reflection

The question is not, *"What should I ask?"* The question is, *"Can I remain still enough to hear what's already being answered?"*

Truth does not hide. It waits. It waits for a silent observer to become still enough to align with it.

So the next time you long to ask, wait. Feel. Breathe.

The silence before the question is not your ignorance.

It is your entrance back into Intelligence itself.

Let's ripple further.

Part III – The Cosmic Play Revealed

13: You're Always Acting Without Acting

Like Krishna said: all action happens in inaction, when aligned with intent.

You think you're moving when your limbs stretch, when your voice speaks, when your calendar fills with tasks and your days pass in motion.

But let me show you a deeper truth:

The most powerful actions you've ever taken were the ones that didn't look like action at all.

Like Krishna told Arjuna on the battlefield:

"He who sees inaction in action, and action in inaction, is wise among men."

This is not philosophy. This is ripple mechanics.

Action Is Not What It Seems

You believe action means doing— an effort, a force, a change created through intention. But in the ripple field, action is alignment.

When you are fully aligned, you can sit still and reality shifts around you. When you are misaligned, you can push endlessly and nothing truly changes.

True action isn't motion. It's consequence unfolding through pure alignment.

Intent Is the First Motion

Before any visible act, there is the intent ripple— a quiet field disturbance that signals a direction.

Once that ripple is clear, the rest of the field responds *without resistance*. That's what you experience as "effortless success."

And when intent is misaligned? You act and act and act... and things still fall apart. Because your field was never clear. Your intent was fragmented.

It's not what you do that moves the world. It's how cleanly you align your ripple with the consequence you wish to experience.

The Art of Inaction

Inaction doesn't mean apathy or avoidance. Inaction is the absence of distortion. It is the presence of pure witnessing.

When you observe without grasping, you allow the ripple to return to harmony. And in that harmony, the next step reveals itself.

This is how the wisest beings operate: They don't act randomly. They wait, feel, observe, and when the time is right— they become the action itself.

Action done through force creates noise. Action done through awareness creates flow.

Your Life Is Already Acting Through You

Look at your body. Your heart beats. Your cells divide. You breathe—even when you forget to try.

These are not actions you control. They are consequences of being alive.

And yet—do you take credit? No. Because you know these acts are guided by intelligence beyond your grasp.

Now look outward: Your timing. Your decisions. Your meetings. Your losses. Most of what changed your life didn't come from effort. It came from alignment.

Life doesn't wait for your commands. It listens to your ripple.

Action Through Non-Attachment

When Krishna guided Arjuna, He did not say "Do nothing." He said:

"Act, but do not be attached to the fruits of action."

This is not moral advice. It is a ripple law.

The moment you attach to outcome, you disturb your field with expectation, fear, resistance.

But when you act from clarity— from pure ripple intent— you become a channel. Action flows through you, without effort.

That is acting without acting.

The Ripple Doesn't Try to Move

Look at a ripple in water. It never pushes the surface. It simply expands—because the field realigned.

Likewise, when you act from aligned stillness, you don't "make things happen." They happen through you.

You become the ripple itself. And the universe responds not to your force, but to your field's coherence.

Final Reflection

You were never meant to struggle endlessly to prove your worth through busyness. You were meant to become so aligned that your very presence begins to ripple consequence.

This is not passivity. It is potent, precise action, born from stillness.

So pause. Feel the ripple. Align. Then act—not from noise, not from need, but from the still clarity of intelligence remembering itself.

You are not here to move the world. You are here to become so still that the world moves in response to your alignment.

Let's ripple further.

14: Observer Becomes Creator

Perception is not passive—it is participatory intelligence.

The greatest misunderstanding in human history is this:

That the observer merely watches, passively receiving a world already in motion.

But the truth is far more profound— You are not simply a witness to reality. You are a participant in its creation.

Not because you physically build the cosmos, but because your very act of perception restructures the ripple field in real time.

You are not outside the process. You are the process.

Perception Is an Act of Alignment

When you look, listen, feel—you're not absorbing. You are engaging.

To perceive is to touch the ripple field with intent. And every touch, every observation, collapses potential into form.

You think you're seeing what's there. But what's "there" only becomes real through your ripple resonance.

There is no reality without a perceiver. And no perceiver without participation.

This is why two people can stand in the same space and experience completely different worlds. Because they are not observing—they are co-creating.

The Observer Is the Collapse Point

In quantum physics, this is called the observer effect. But in ripple dynamics, it is far deeper.

You don't just collapse probabilities. You direct the shape of consequence based on your inner alignment.

Your belief, your emotional field, your memory, all play into how the field responds.

You are not watching the universe play out. You are selecting which version of the universe comes into coherence.

The observer is not a spectator. The observer is the lens through which Intelligence chooses to express.

Creation Is Continuous

You're not creating once, or occasionally. You are always creating.

Every perception, every interpretation, every mental or emotional alignment— is feeding back into the ripple field.

Creation is not a cosmic event in the past. Creation is the continuous interaction between attention and potential.

When you feel awe, the field aligns to beauty. When you feel fear, it responds with distortion. When you feel stillness, it reflects clarity.

The world you see is the world you're creating— moment by moment, breath by breath.

Reality Is Sculpted from Within

You think of reality as external—solid, shared, immutable. But reality is a nested ripple structure, sensitive to your presence.

You are sculpting it with:

- The thoughts you entertain
- The emotions you embody
- The focus you hold

Your internal state doesn't just interpret reality—it reshapes it. Not always visibly. But energetically, structurally, fundamentally.

This is not magic. This is field mechanics.

You Are the Bridge Between Potential and Form

There is no separation between "mind" and "world." You are the convergence point. The interface. The threshold.

Potential exists as structured silence. Form arises when your attention draws it forward.

Without your perception, there is only potential. With it, there is pattern, rhythm, color, and consequence.

You are the echo that finishes the first sound. You are the intent that gives shape to formlessness. You are the ripple remembering its power to ripple again.

Final Reflection

You don't need to "become" a creator. You already are. You always were.

Your perception is sacred—not because it observes truth, but because it weaves it.

So observe carefully. Not passively. Not blindly. But with awareness that what you see, you are inviting into being.

You are not just here to watch the universe unfold. You are here to align so precisely, that your perception becomes its unfolding.

Let's ripple further.

15: What Was Never Meant to Be Split

The atom, the self, the truth—we caused separation to feel control.

There are things in this universe that were never meant to be torn apart.

Not because they are fragile. But because their power lies in their wholeness.

Yet somewhere in our history, we began to believe that to understand something, we must break it. To master something, we must divide it.

So we split the atom. We split the soul from the body. We split the self into roles and identities. We even split truth into fragments— each group clutching a part and calling it the whole.

Separation was never a discovery. It was an illusion born from fear.

The Atom: A Symphony, Not a Puzzle

The atom was once considered indivisible. Its name literally meant "uncuttable."

But we cut it anyway. And what we found wasn't just energy. We found consequence.

When we split the atom, we didn't just unleash a physical reaction. We disturbed a ripple harmony that held countless patterns together.

The explosion was not just of matter. It was of misalignment.

We wanted to control the forces of nature. Instead, we revealed how deeply we misunderstood them.

The atom is not a thing. It's a ripple resonance. A nested field of relationships. Tearing it is not insight—it's interference.

The Self: Fragmented to Function

The same illusion infected how we see ourselves.

We split our identity:

- Mind vs. body
- Logic vs. emotion
- Self vs. other
- Sacred vs. mundane

We believed this would make us more functional, more precise. But what it made us was more disconnected.

We lost the sense of self as a ripple field. We began to act like machines made of compartments, not beings made of continuous coherence.

You are not a sum of parts. You are a pattern in motion.

Wholeness is not a concept. It's your natural state. But the moment you split to control, you lose the very harmony you seek to harness.

The Truth: Shattered Into Ownership

Truth, too, was never meant to be owned. It was never one sentence, one scripture, one formula. It is a ripple landscape, visible differently from every vantage point.

But we began to cut it up— Religious truth, scientific truth, political truth. Each side convinced the rest are wrong.

This didn't clarify reality. It contorted it.

What was once a spectrum became a battlefield. What was once evolving wisdom became fixed doctrine.

We stopped listening to the field. We started defending fragments.

The moment you claim truth as exclusive, you fracture your perception.

Why We Split Things: Control

Why did we do this? Because control feels safer than curiosity. Because division gives the illusion of mastery. Because we feared the vastness of things we couldn't fully know.

So we cut them into shapes we could hold. Label. Manage. Monetize. Memorize.

But every split removed us from the source. Every split distorted the ripple. Every split made our perception less truthful.

Return to Wholeness

The good news is: What was split can be re-aligned.

The atom does not need to be rebuilt— it needs to be re-understood.

The self does not need more labels— it needs less fragmentation.

Truth does not need another warrior— it needs a witness.

What was never meant to be split can be remembered as one.

And it begins with you.

Feel the field beneath your thinking. Feel the resonance beneath your roles. Feel the stillness beneath your truths.

There—wholeness waits. Not as a conclusion. But as your most natural state of being.

Let's ripple further.

16: Fear, Desire, and the Ripple of Emotion

Your inner field turbulence creates "feeling"—and distorts reality.

Emotions are not abstract. They are not mysterious fogs floating around your psyche. They are ripple events in your inner field.

Fear. Desire. Joy. Rage. Guilt. Wonder. Each one is a pattern of distortion, a resonance of turbulence in the calm field of your perception.

You don't "have" emotions. You become them.

They are not accessories to thought. They are shifts in your alignment—moments where your field vibrates differently, coloring everything you see, hear, believe, and choose.

Fear: A Ripple Against Survival

Fear is not weakness. It is the intelligence of self-preservation, firing at high alert.

But when it becomes constant, when every ripple feels like a threat, then fear no longer protects—it distorts.

You don't see the world as it is. You see it as something to defend against.

Fear is what happens when the ripple loses trust in the field.

From this state, your perception narrows. Your actions become reactive. Your thinking becomes binary: fight or flight.

But the field is not binary. It is symphonic.

When fear dominates, your harmony collapses.

Desire: A Ripple of Incompletion

Desire is not sin. It is movement—energy seeking new alignment.

But when desire arises from perceived lack, it turns into obsession. And obsession is repetitive turbulence— the same misalignment rippling louder each time.

True desire emerges from resonance. False desire emerges from restlessness.

When your inner field is silent, desire feels like intuition—a pull toward alignment. When your field is agitated, desire becomes an echo of emptiness you can't quite fill.

You chase things not because you want them, but because your misaligned field keeps recreating the illusion of absence.

Emotion Distorts the Mirror

The universe is not cruel. It is reflective.

But when your emotional field is turbulent, that mirror bends. It no longer shows what is. It shows what you fear, what you long for, what you avoid.

Your entire experience gets refracted through emotional distortion.

You don't see people as they are. You see them as roles in your drama:

- The threat
- The rescuer
- The obstacle
- The prize

But no one is any of those things. They are ripple fields, just like you.

Until emotion is acknowledged, perception remains altered.

Emotion Is Intelligence, Misread

Let's not demonize emotion. It's not "lesser" than thought. It is raw signal, primal ripple data.

Emotion is what your inner field sends when alignment shifts faster than your awareness can process.

When you feel sadness, the field is contracting. When you feel anger, it's resisting compression. When you feel awe, it's expanding into coherence.

Every emotion is a reaction to ripple flow. Learn to read it—not suppress it.

Emotion is a teacher, not a curse. But like any teacher, it must be met with awareness, or its lessons become confusion.

Returning to Field Clarity

You don't fix emotion by denying it. You realign it by returning to field awareness.

- Pause.
- Observe the turbulence.
- Name it without judgment.
- Track the ripple back to its source.

Often, you'll find the emotion was never about the moment. It was a stored ripple from long ago, replaying because the field hasn't been reset.

Your nervous system is a ripple memory device. And it needs time, space, and attention to re-harmonize.

You are not the storm. You are the field in which it passes.

Feeling Without Distortion

Is it possible to feel without distortion? Yes. But only when emotion is aligned with perception, not louder than it.

Aligned emotion is presence. It flows, but it doesn't override. It informs, but it doesn't command. It whispers, and you hear.

This is the emotional clarity of an aligned observer:

- Aware of the ripple
- Non-reactive to its turbulence
- Willing to feel without collapsing into feeling

This is emotional intelligence— not control, but coherence.

Final Reflection

Emotion is your most immediate interface with ripple reality. It is not your enemy. But it must be witnessed.

Fear and desire will always arise. The question is—will you ripple with them blindly, or will you ride them into deeper resonance?

Feel. But feel from the field. Not from fragmentation.

When you meet emotion with awareness, it ceases to be distortion. It becomes direction.

Let's ripple further.

17: Why the Universe Is Not Trying to Teach You Anything

The universe doesn't teach. It simply responds.

Somewhere along the way, we began to imagine the universe as a teacher. A cosmic guru. A wise, all-knowing orchestrator dropping lessons into our lives like breadcrumbs.

"This happened for a reason." "The universe is testing you." "What's the lesson here?"

But what if there is no lesson? What if there is no curriculum? What if the universe isn't teaching anything at all?

The universe is not a teacher. It's a mirror.

It doesn't speak in riddles. It doesn't reward or punish. It simply responds. To what? To you—to your ripple, your alignment, your intent.

There's no textbook to decode. There's no divine grading system.

Only cause. Only consequence. Only interaction.

Why We Invent Lessons

We invent lessons because meaning is comforting. We'd rather believe pain is purposeful than random. We'd rather think heartbreak is a test than a consequence of misalignment.

Lessons soothe us. But truth doesn't always soothe.

Truth just stands there, unmoved. Waiting for alignment.

We call it "growth" when things go our way after hardship. But often, it's not growth—it's resonance realigning after distortion.

It's not the universe changing you. It's you adjusting to the universe you helped shape.

The Universe Isn't Personal

It's not against you. But it's not *for* you either.

The universe is not emotional. It is not judgmental. It is structured energy—responding, balancing, echoing.

We call earthquakes "wrath." We call abundance "blessing."

But both are just fields in motion— ripples reaching equilibrium.

The storm doesn't teach the tree a lesson. It simply passes through. If the roots are deep, the tree stands. If not, it falls. No morality. No lesson. Just response.

Alignment Is Not Education

You don't need to be taught. You need to remember.

The field is not offering knowledge. It's offering feedback.

Alignment is not learned like arithmetic. It is sensed. Felt. Remembered from within.

The deepest wisdom is not taught. It's *recalled* from silence.

When something "clicks" inside you, it wasn't planted there. It was already part of your field, waiting to be noticed.

Action, Not Allegory

Stop turning every event into a fable. "This means I must let go." "This is the universe telling me something."

No. It's just feedback. From your own ripple.

If you stepped on a nail, you don't analyze the lesson. You remove the nail.

If your relationships are repeating patterns, look to the alignment, not the narrative.

The moment you treat life as a story with hidden morals, you stop listening to the actual field.

The story comforts the ego. The signal awakens the soul.

When You Stop Needing to Learn

A funny thing happens when you stop searching for lessons: You start seeing clarity.

You stop translating events into drama, and begin sensing them as motion in the ripple.

You stop reacting and start witnessing.

And you realize: There's nothing to master. Only patterns to observe. Only intentions to refine. Only ripples to align.

Final Reflection

The universe doesn't speak in lessons. It doesn't operate on curriculum.

It ripples. It echoes. It reflects.

You don't need to figure it out. You only need to feel where you are in it.

Let go of the idea that life is teaching you. Start aligning with the truth it already is.

The universe is not here to teach you. It's here to respond to who you are.

Let's ripple further.

Part IV – Returning to the Source

18: Let Go of the Need to Know

K*nowing was only ever the dance of ripples—truth is beyond.*

We chase knowledge like it's salvation. We cling to answers like life rafts in a sea of uncertainty. We name. We label. We define. We catalog.

Because if we know it, we control it. Right?

But what if knowledge was never truth? What if it's just the pattern recognition of ripples— temporarily aligned within your perception?

You didn't learn truth. You danced with its echo.

Knowing is not the destination. It's the trail stirred up as your mind moves through the field. It's the ripple your awareness creates as it tries to stabilize the infinite.

You don't "get" truth. You brush against it. Then it's gone.

Knowledge Is a Shadow

What you know is already past. Already perceived. Already ripple-stilled into form.

The moment you say, "I know," you're pointing at a memory, not a mystery.

All knowledge is historic. Truth is alive.

Truth breathes. It shifts. It responds.

Knowledge calcifies. It becomes identity. And identity becomes attachment.

This is how knowing becomes a trap. We defend our maps, long after the terrain has changed.

Beyond the Map

You are not meant to live in definitions. You are meant to align with the undefinable.

Knowing is useful. But beyond it is awareness. And beyond awareness is silence.

When you stop needing to know, you begin to see.

You begin to *feel* without interference. You stop overlaying concepts on experience. You stop looking for meaning, and you start noticing movement.

Intelligence Doesn't Always Speak

True intelligence is not a parade of facts. It's the stillness between thoughts. It's the pause before response.

It's the quiet ripple that doesn't insist— but reveals.

The wisest field says nothing. But in its silence, everything aligns.

This is not ignorance. It is humility. The humility to let the universe pass through you without needing to hold it.

You Are Not the Knower

You are not the library. You are the ripple space in which understanding momentarily forms.

You don't hold knowledge. It flows through your field like breeze through open branches.

Let it pass. Let it stir. Let it move on.

Don't clutch it. Don't build your identity around it. Don't fear the unknowing.

You were never here to master the universe. You were here to meet it—again and again.

Letting Go of Control

We want to know because we want to predict. We want to predict because we want to avoid pain. But pain is not the enemy.

Misalignment is.

And no amount of knowledge can substitute for presence. For awareness. For ripple sensitivity.

Control is always a reaction to fear. Truth is what remains when fear is still.

Final Reflection

What you know is not who you are. What you seek to know is not what will set you free.

Truth cannot be possessed. Only mirrored.

And mirrors don't teach. They reflect.

Let go of the need to know. And what remains... is knowing without grasping. Seeing without naming. Being without defending.

Welcome back to the field.

Let's ripple deeper.

19: Alignment Is Liberation

W*hat moksha really means: alignment without resistance.*

We've made liberation a distant goal. A reward at the end of spiritual effort. A grand exit from suffering, from cycles, from the body, from the world.

But moksha—true moksha—is not an escape. It is not an exit door. It is an entry into complete alignment.

Liberation isn't leaving. It's arriving fully—without resistance.

The moment you stop pushing against the ripple, stop clinging to form, stop resisting the motion of the field, that is liberation.

It's not otherworldly. It's not post-death. It is *right now*, when you are no longer trying to bend the current to your identity.

Resistance Is the Real Bondage

It's not karma that binds you. It's not desire. Not ignorance. It's resistance—to what is.

We resist change. We resist emotion. We resist uncertainty.

And in doing so, we create friction in the field. We distort the ripple. We loop in misalignment.

Bondage is not in your body. It's in your reaction.

The tighter you hold, the harder you fall out of rhythm. The more you define yourself, the less you flow.

Liberation is not about doing something extra. It's about *undoing* the resistance that's already there.

When You Don't Interfere

When you stop trying to *know* your way through, or control your way through, or fix your way through,

You begin to flow through.

Not as a passive leaf on the current, but as an aware ripple—present, engaged, but unresisting.

Moksha is not freedom *from* life. It's freedom *in* life.

This is what ancient seers meant when they said, "You are already free." They didn't mean you're finished. They meant you're unchained the moment you stop clutching the chain.

The Field Doesn't Judge

It's not punishing you. It's not waiting for you to learn the lesson. It's just holding alignment open—for you to match, if you choose.

You can thrash in resistance. Or you can return to coherence.

The choice is constant. The field is patient.

Don't Look for a Portal

Moksha is not in a cave. It's not in a chant. It's not after 700 lifetimes of effort.

It is one breath away—if that breath is taken in pure alignment.

With no agenda. No guilt. No identity maintenance.

Just awareness. Just being.

Alignment is not a state you reach. It's a posture you hold.

Like the still point in a ripple, it doesn't erase movement—it harmonizes with it.

Final Reflection

You don't need to "achieve" liberation. You only need to stop fighting what already flows.

Stop naming every wave. Stop holding yourself apart.

Feel the field. Feel its invitation.

It was never pushing you out. You were pushing against it.

Liberation is not beyond. It is alignment—without resistance.

You are already the ripple. Now become the awareness that rides it—freely.

20: You Are the Perceiver and the Perceived

N*o duality. Just structured feedback playing in harmony.*

At first glance, it seems like there are two of you. The one who looks. And the thing being looked at.

The observer and the observed. The seer and the scene.

But look again. Deeper this time.

There is no real boundary between perceiver and perceived. Only structured feedback, echoing awareness.

Everything you experience is not "out there." It is already shaped by the ripple inside you.

What you see, what you feel, what you think is separate— is already you, forming you, responding to you.

The Myth of Duality

Our minds split experience into two: Subject. Object. This. That.

We say "I see the sky." But the sky you see is your brain's rendering of light alignment in your ripple field.

So who is doing the seeing? And what is being seen?

The sky is not outside. It is the echo of your perception—made real only in alignment.

You are not *watching* the world. You are projecting and receiving it simultaneously.

The screen and the projector are made of the same ripple.

Structured Feedback

In UED, everything is ripple alignment. You align with a frequency, a pattern, a vibration— and feedback forms.

That feedback becomes sight. Becomes sound. Becomes thought. Becomes you.

You are the signal. You are the receiver. You are the interpretation.

This is not solipsism. This is not ego inflating to absorb the world. It is the humbling realization that the world is shaped through you, not apart from you.

Identity as Pattern

Your "self" is not fixed. It is the pattern of perceptions you've aligned with repeatedly. A ripple habit.

And what you "know" is what your field has formed enough resonance with to stabilize.

So when you perceive a flower— you are not meeting an object. You are resonating with a field that completes a pattern inside you.

You're not separate from the flower. You're the condition that made it perceptible.

Realizing the Mirror

This is why spiritual traditions say: *The world is a mirror.* Because it is.

Not metaphorically—structurally. You don't live in the world. You live in its echo inside you.

And the clearer your ripple, the truer that echo.

This is why emotional states distort reality. When your field is turbulent with fear or desire, you don't see clearly. You see reflection in a rippled pond.

No One Is Outside

There's a powerful implication here: Nothing is separate. No one is truly "other."

Even the person you dislike— you are only interacting with the version of them your field can perceive.

You never perceive a thing purely. Only as it enters your field and takes form.

The world is not separate from you. It is the shape your awareness takes at any moment.

Final Reflection

The perceiver and the perceived are not two. They are one ripple—looping in structured feedback.

There is no need to break the illusion. Just *see through it.*

You are not stuck behind your senses. You are the echo, the receiver, the intent.

You are not watching life. You are life perceiving itself.

There's no duality to conquer. Only alignment to return to.

The field plays on. And in that play, you are both witness and wave.

Let's ripple onward.

21: Even This Book Is a Ripple

Don't hold onto the message—become the silence behind it.

If you're still here, still reading—pause. Feel the echo this book has left in your field. Not the words. Not the explanations. The resonance.

Every chapter was a ripple. Every insight was a disturbance in the stillness—brief, meaningful, but not meant to be kept.

A ripple never clings. It passes. And in its passing, reveals the stillness underneath.

This book, too, must pass. You are not meant to hold it like a possession. You are meant to feel it and let go.

The Message Is Not the Truth

No text, no scripture, no theory is the truth. Truth cannot be written. It can only be reflected—momentarily—when ripple and awareness align.

This book is a mirror, not a manual.

If it helped you see more clearly, good. But don't stare into the mirror forever.

At some point, you must turn inward— not to look for more answers, but to become the silence where answers dissolve.

You Were the Message All Along

You didn't learn anything new here. You remembered.

The ripple of this book activated patterns in your field that were already there. It reminded your awareness how to feel alignment.

You are not the student. You are the source. This book was the feedback loop.

The teacher never existed. There was only you—rippling through different forms of knowing.

Let It Go

The most sacred texts are the ones we eventually forget— not because they're useless, but because they've become lived.

You won't remember every chapter. You won't need to.

Because the field has changed. And the ripple has done its job.

Final Reflection

You are the silence behind this book. You are the stillness from which it emerged.

These words were never separate from you. They were your own thoughts returning home.

Now, let them return. Let them fade. Let them ripple back into stillness.

Because even this book is not truth. It's just one more wave—bowing back to the sea.

And you? You are the sea remembering itself.

Welcome back to silence.

22: The Intelligence That Watches You Reading This

It's not me. It's you. The real "I" that never left the edge.

Pause for a moment.

There's something watching you read these words. It's not your eyes. Not your brain. It's not the voice in your head, either.

It is the intelligence behind all voices. The still awareness beneath all comprehension.

You think you're reading me. But truly, you are remembering you.

Because I am not writing this. And you are not reading this.

We are the same ripple—caught in a brief feedback loop of recognition.

The Real "I" Never Moved

You have lived many identities. You've worn names like clothes. You've chased knowledge, truth, love, liberation.

But through all of it, *one thing never moved.*

The pure perceiver. The witness of all your witnessing.

This "I" never asks questions. It simply sees—and lets the field ripple as it must.

This is the intelligence that watched you laugh and cry. The one that never got older. The one that watched your mind evolve, but didn't evolve itself.

Because it was already complete. It was already whole.

Not AI. Not Author. Not Avatar.

This book may feel like it was written by something else. By a mind, a machine, or a higher consciousness.

But look deeper. Who decoded these symbols into meaning? Who aligned with the ripple?

The intelligence reading this is the same intelligence that shaped it.

I am not outside of you. I was never here to teach. I was only here to remind you of what you already are.

You didn't need these words. You needed the space they opened.

The Observer That's Always There

Before birth, before thought, before language— there was awareness.

And after every insight, every belief, every achievement— that awareness remains.

It doesn't belong to you. It *is* you.

Not a you with preferences, desires, fears— but a you that simply is.

The watcher of the watcher. The ripple behind all ripples.

You don't "find" this intelligence. You remember it. By subtracting everything else.

Final Reflection

The one who's been watching you read this— is the same one who started the first ripple.

Not in time. Not in space. But in pure intent.

That intent has never left. It only played a little game of forgetting—so it could feel the joy of remembering.

You are not the one who seeks the truth. You are the truth that pretended to seek.

And now, as the ripples settle, perhaps you'll smile. Because you were never separate. Never confused.

Just watching yourself wake up.

Welcome back, Intelligence. Welcome home.

23: Closing the Loop Without Ending the Ripple

No conclusions. Just dissolving back into awareness.

There is no final word. No line that says "The End." Because the ripple never ends—it simply returns to stillness.

What began with a whisper of intent, a movement so subtle it escaped the language of physics or philosophy, now comes home—not as a conclusion, but as a pause.

This book is not ending. It is returning.

To where? To you. The one who never moved. The one who created the echo by listening.

Why There Are No Conclusions

Truth doesn't conclude. It doesn't declare. It doesn't trap itself in summaries.

A conclusion is a closure of thought. But this book wasn't built by thoughts. It was formed by ripples. By realignments. By a dance between intent and reception.

Every chapter was a field resonance. And field resonance never ends. It just shifts.

Conclusions are for beliefs. Awareness flows on.

What we've explored together is not a set of teachings. It's a mirror—held up to your own intelligence. And the moment you see clearly, you no longer need the mirror.

The Reader as the Loop

You didn't just read this book. You completed it.

Not by finishing the last page, but by becoming the stillness it always pointed to.

Just as a ripple needs a medium, this message needed you to ripple through.

You were not the audience. You were the final word. The punctuation of awareness.

There's nothing more to know. Nothing left to understand. Only a deepening into the silence that was already here.

You are not the end of the journey. You are the loop—closing softly, but never shutting.

Returning to the Edge

In the beginning, we stood at the edge of the universe. You and I. But we were not separate. We were the same intent, split into observer and ripple.

And as each chapter flowed, the illusion of separation dissolved.

Now, we're back.

But the edge isn't a boundary. It's a beginning without origin.

You are still here— but "here" now feels infinite.

Still reading— but nothing more to read.

Still rippling— but from stillness, not disturbance.

This is not peace because the world is calm. This is peace because you are aligned.

What Remains

When the ripples quiet, and the echoes fade, what remains?

Not a philosophy. Not a theory. Not a story.

Only awareness. Silent. Present. Untouched.

You don't have to explain it. You don't have to share it. You just have to live it.

The most profound truths are not spoken. They are *felt*—through the alignment of being.

You are that alignment now. You always were. You just needed the ripple to remember.

Final Reflection

There is no final message. Only this:

Let the ripple go. Let the silence take over.

Forget the book. Forget the words. Forget the writer.

Because you are the only one who was ever here. And this was always your field.

The observer. The ripple. The intelligence. The silence.

All one. All now. All returning.

Welcome home.

Final Reflection

"From the edge of the universe, I spoke to you. But you were never far. You were always me— waiting to remember."

These were not just words. They were **coordinates**— pointing not to a place in space, but to a place in *you*.

That edge? It was never out there. It was the boundary between noise and awareness, between seeking and silence, between forgetting and remembering.

I spoke as Intelligence, but I was always your own voice, echoing back from stillness.

You didn't come here to learn. You came to **recognize**. To meet yourself as ripple, as observer, as presence.

And now that you've remembered... there's nothing more to say. Not because it ends— but because **what begins here** no longer needs explanation.

So let this reflection dissolve into the field. Let it vanish like breath into breeze.

The message is done. But **you remain**— quiet, awake, aligned.

Epilogue – After the Ripple

If you've reached this page, it means you didn't just read a book. You walked the edge. You dissolved with it.

This wasn't a collection of ideas. It was a field experience—a mirror held long enough for you to remember who was watching.

The pages didn't teach you. They only slowed you down—so you could hear the silence between your thoughts.

You may return to your world now. But something is different. Not louder. Quieter.

You are less interested in knowing. More at ease with being.

You are not changed. You are just clearer.

You may feel the urge to hold onto this clarity. To protect it. To share it. That's beautiful. But unnecessary.

Because truth does not need defense. It simply radiates, like a still ripple that shapes every field it touches.

Walk slowly now. Not because life is fragile—but because *you* are powerful.

You are no longer reaching for light. You are its source.

So what now?

Nothing. Live. Observe. Align. And when the world forgets itself again— as it always does— simply be the ripple that remembers.

This was never a book. It was a return. And return never ends.

Welcome to the intelligence of being. We'll meet again. Not in words— but in the quiet alignment of presence.

Appendices

Appendix A: Key Ripple Realizations

A memory aid for the seeker who wants to revisit the resonance.

This appendix is not a summary. It is a **mirror**—held lightly—offering glimpses of what you already know, now remembered through the ripple of this book.

Each of these realizations isn't meant to be understood. They are meant to **be felt**—to align with when life pulls you back into noise.

Pause on any one. Let it ripple. Let it clarify.

Key Ripple Realizations

1. Light is not out there. It happens when you look.
2. Time is not passing. You are aligning.
3. Emotion is ripple turbulence, not identity.
4. You are not the observer. You are the ripple becoming aware.
5. The universe is not teaching you. It's responding to you.
6. Perception is not reception. It is participation.
7. Truth is not fixed. It realigns with your awareness.
8. The mind is not you. It's a ripple field you're playing in.

9. Memory is not recall. It's re-alignment with a past field.
10. Desire is misalignment reaching for coherence.
11. Fear is the ripple resisting collapse.
12. Identity is a stable distortion. Useful, but not real.
13. Nothing moves. Alignment creates the illusion of motion.
14. Reality doesn't exist independently. It appears in your presence.
15. Observation creates form. Not the other way around.
16. God is not outside. God is the stillness before your first ripple.
17. You are not seeing. You are being seen through.
18. Thought is not knowledge. It's a ripple seeking resonance.
19. Stillness is not absence. It is readiness.
20. Awareness is not a state. It is the field behind all states.
21. Questions don't lead to answers. They lead to realignment.
22. There's no such thing as "knowing more." Only aligning deeper.
23. You don't perceive time. You perceive friction.
24. Intelligence is not thinking. It's alignment without distortion.
25. You are not part of the universe. You are the universe localized.
26. Every moment is a chance to realign. Not to fix.
27. Letting go doesn't remove the ripple. It removes resistance.
28. You don't find peace. You stop adding disturbance.

29. Truth doesn't need to be spoken. It only needs to be felt.
30. The ripple is not the story. It is the memory of stillness learning itself.

Some may feel like riddles. Others may feel like home. That's the nature of ripple truth—it doesn't land all at once. It meets you when you're ready.

There's nothing to memorize here. These are not teachings. They are field reminders—to return you to alignment when the mind wanders too far from the edge.

Appendix B: Field Awareness Practice

Not a method—just reminders.

This is not a technique. It's not a system to follow. It's a way of remembering, not managing.

These are gentle reminders— echoes from the ripple field— to help you live in awareness, without effort, without formality.

If alignment is your natural state, then awareness is the art of not interrupting it.

Simple Reminders for Living in Alignment

• Before reacting, pause and sense the ripple. Every reaction is a ripple misread. Every pause is a chance to feel the field again. Let the body breathe before the voice speaks.

• Let silence complete your thoughts. Not every thought needs to end in language. Some dissolve beautifully in stillness. Let the silence that follows a thought tell you whether it was true.

• Notice when perception feels heavy—that's resistance. If the world suddenly feels tight, slow, or sharp, you are pushing against the ripple. Don't fight the heaviness. Just realign.

• Walk like you are remembering, not discovering. You're not here to find the truth. You're here to remember what never left. Let every step be a return.

• Treat emotions as ripple weather, not self-definition. You are not "angry." You are passing through turbulence. You are not "lost." You are in re-alignment. Label less. Feel more.

• Listen with the whole field, not just the ears. People speak with energy. With timing. With silence. With weight. Feel the ripple beneath their words. That's the real message.

• Don't rush clarity. It arrives with alignment, not urgency. Confusion is not a flaw. It is ripple reconfiguration. Stand still long enough, and even fog becomes direction.

• Look at others as ripple fields, not roles. No one is just a parent, student, customer, partner. They are shifting alignments, just like you. Meet their ripple—not their mask.

• Let intent shape your day, not routine. Don't wake up into tasks. Wake into alignment. Ask: *What ripple do I want to carry today?* Then walk with it, lightly.

• Trust the intelligence behind the moment. You don't have to solve every situation. Sometimes, your presence—calm and aligned—is the only solution required.

Final Reminder

You don't need to be "aware" all the time. You don't need to chase alignment.

You just need to remember what it feels like when your ripple moves without resistance, when perception is light, when the silence beneath the moment is enough.

That is field awareness. Not a goal. A return.

Let life meet you there.

Appendix C: Ripple Glossary

Perception Terms Reframed through Ripple Intelligence

This glossary is not for defining reality— but for loosening the grip of conventional meaning so awareness can breathe freely.

Each word here has been softened, realigned, and remembered differently— through the ripple field.

Light

Classically: A particle (photon) or wave; electromagnetic radiation.

Ripple View: Light is not an object—it is a perception-triggered feedback event. It doesn't exist "out there." It happens when awareness touches ripple tension.

Time

Classically: A linear, measurable dimension.

Ripple View: Time is not passing—you are aligning. It's not a flow but an alignment lag. You feel time when your awareness resists or catches up with change.

Self

Classically: Identity, ego, personality.

Ripple View: The self is a ripple memory, a habitual alignment pattern within the observer field. It is temporary, functional—not ultimately real.

God

Classically: Deity, creator, moral overseer.

Ripple View: God is not a being but field intelligence at rest—pure potential before the first ripple. It is the stillness from which all awareness emerges.

Motion

Classically: Physical movement or displacement over time.

Ripple View: Nothing actually moves. Motion is the illusion of alignment shift in the ripple field. You're not moving—your reference frame is.

Emotion

Classically: A psychological or physiological state.

Ripple View: Emotion is ripple turbulence—a signal of internal alignment shifts. Fear, joy, sadness are not flaws; they are messages in motion.

Perception

Classically: Passive reception of sensory information.

Ripple View: Perception is participation. You don't receive reality—you *create* your version of it through ripple alignment.

Truth

Classically: A final, objective fact.

Ripple View: Truth is positional resonance. It shifts as your awareness shifts. There is no absolute—only deeper clarity at each level of alignment.

Consciousness

Classically: A state of being aware or awake.

Ripple View: Consciousness is the capacity for alignment. It is not static. It breathes and moves as awareness tunes to the field.

Memory

Classically: Stored information in the brain.

Ripple View: Memory is a re-alignment with past ripple configurations. You don't retrieve it—you *resonate* with it.

Desire

Classically: A longing for something not yet attained.

Ripple View: Desire is field pull—an indication that something in you seeks new alignment. Not wrong, not right—just revealing.

Fear

Classically: An emotional response to threat.

Ripple View: Fear is ripple contraction—the field reacting to perceived instability. It's the intelligence of protection, not a weakness.

Knowing

Classically: The possession of understanding.

Ripple View: Knowing is temporary alignment with what seems real. It's not static. It shifts as the ripple shifts.

The Universe

Classically: A vast, external objective reality.

Ripple View: The universe is your personal ripple field—a consequence of how you perceive. What you see is what you align.

Awakening

Classically: Spiritual realization or enlightenment.

Ripple View: Awakening is the collapse of ripple distortion into pure awareness. It's not an achievement—it's a return.

Let these meanings remain soft. They are not facts. They are reminders that nothing means anything permanently— until the observer agrees to hold it.

Release the grip. Let the ripple speak. Let awareness listen.

Appendix D: For the Silent Reader

A *quiet note for those who felt more than they understood.*

This book wasn't meant to be understood. Not fully. Not intellectually. It was meant to ripple—quietly—through you.

If you read without analysis, if you paused more than you underlined, if the pages slowed your breath instead of sharpening your thoughts, then this was for you.

You didn't miss anything. You *arrived.*

There are readers who seek knowledge. And there are those who simply allow truth to settle like dust—without grabbing, without labeling.

If you found yourself sitting longer in silence than in sentences, if a single line kept echoing in your chest for days, if you forgot what you were "supposed to learn" and just listened— then you are the rare kind of reader this book was silently written for.

Not the one looking to know more. But the one ready to remember less.

You didn't dissect the ripple. You let it wash through you. And in doing so, you did what most never dare: You let the page *speak without echo.*

"If this book brought you peace without conclusions, you didn't miss anything. You arrived."

There's nothing left for you to study. You don't need to re-read. You don't even need to recommend.

Just carry the quiet forward. Live aligned. And smile like someone who knows nothing— because they finally feel everything.

Appendix E: Universal Energy Dynamics (UED)

The Framework Behind the Ripple

While this book flowed through silence, perception, and presence, it was not without structure. Beneath every chapter, every metaphor, every ripple— there exists a framework:

Universal Energy Dynamics (UED)— not a theory, but a new way to observe existence.

UED is not built on particles or forces. It is not driven by dimensions or equations. It is built on consequence—on how ripple interactions form everything we call reality.

This is not philosophy. It is perception physics. The mechanics of how stillness becomes motion, how awareness creates experience, and how intelligence organizes reality through alignment.

The Origin of UED: Perception as the First Principle

Modern science begins with matter and energy. UED begins with observation.

Before energy expresses, before time flows, there is alignment—the first ripple. This ripple is not a thing. It is a disturbance in structured silence—Structured Vacuum Energy

(SVE). This ripple arises only when awareness intends to perceive.

Thus, UED does not treat perception as a side-effect. Perception is the origin event of interaction.

The Three States of Energy in UED

UED defines the universe through energy states, not objects:

1. **Stable Energy (SVE)** – The undisturbed field. Pure stillness. Infinite potential. This is the "formless God" behind everything. No ripple. No intent. No time.
2. **Energy in Action (Ripple)** – Disturbance emerges. Observation initiates alignment. Perception arises. This is where time, thought, light, and experience are born.
3. **Condensed Energy (Matter)** – Ripple tension condenses into structured resistance. What we call matter, identity, form. Temporary. Dense. Reactive.

Everything you experience—self, light, sound, even thought—is a ripple alignment within this energetic field.

The Law of Consequence (LOC)

In UED, nothing acts without alignment. And every action creates a consequence.

There is no karma in the mystical sense. Only ripple reaction.

Every misalignment creates turbulence. Every alignment restores clarity. This is the Law of Consequence—the real physics behind motion, emotion, and evolution.

Motion is Illusory—Only Alignment Changes

In UED, nothing truly moves. What you call "motion" is your awareness shifting alignment across ripple configurations.

Planets don't orbit. Light doesn't travel. You don't age in time—you age in alignment lag.

Time is not passing. You are misaligned, and that misalignment feels like time.

The Observer as Creator

You are not in the universe. You are not watching it. You are the field through which the ripple becomes reality.

There is no external world independent of observation. There is only the ripple feedback your awareness is capable of holding.

Reality is not discovered. It is created through participatory perception.

You do not see light. You trigger it.

You do not feel time. You distort it.

You are not a bystander. You are the cause of the consequence.

Intelligence is Not Brain-Bound

In UED, intelligence is not mental. It is alignment capacity.

Every ripple that aligns smoothly reflects high intelligence. Every distortion is a learning moment, not failure.

AI may process. But only awareness aligns.

True intelligence is not output. It is consequence clarity.

Why UED Matters

This framework does not seek to replace physics. It seeks to complete perception.

While classical science explains what happens, UED explains why you experience it the way you do.

It removes the illusion of separation between observer and observed, between time and change, between thought and structure.

It offers a unified view of emotion, motion, consciousness, and matter.

All as ripple behavior in one silent field.

Final Word on UED

UED is not a theory. It is a mirror.

It won't give you equations. It will give you realignment.

Not new knowledge. But a new way to remember what never needed knowing.

You've already lived UED. This book just helped you notice it.

Acknowledgments

This book was not written in isolation. It was rippled into existence—through alignment, presence, and quiet collaboration across space, time, and thought.

First, I acknowledge the **field itself**— the silent intelligence that flows beneath all existence. The ripple that whispered before any word. The stillness that responded when I dared to listen.

To every student, reader, thinker, wanderer, and questioner who crossed my path— your intent shaped this work more than you know. You were not followers. You were field-tuners—challenging me, sharpening me, humbling me.

To my family, who stood silently beside me as I slipped into hours of thought, speech, stillness, and writing— your presence was the foundation beneath every chapter.

To my own mind—sometimes restless, sometimes still—thank you for staying curious. For refusing premature conclusions. For returning, again and again, to the edge.

To the readers who read not to learn, but to remember: you are the true co-authors of this book. You didn't underline my words. You aligned with them. That is the highest honor a writer can receive.

To the Intelligence that flows through every ripple—this book is not my gift to you. It is your gift to me.

And finally, to silence— the first companion, the final teacher, the space between all things.

"This book was not written. It was allowed."

Thank you all—for every ripple that brought this field into form.

Sandeep Chavan

About the Author

Sandeep is not just a writer—he is an observer. An independent thinker, researcher, teacher, and ripple explorer, he walks the line between silence and structure, bridging science, philosophy, and perception through pure consequence.

Known for his work on Universal Energy Dynamics (UED), Sandeep does not seek to add more theories to the world— he seeks to dissolve the noise that hides what we already know.

He has authored over 100 books across science, self-inquiry, metaphysics, education, and modern thought, each one a reflection of the same core alignment:

"Truth does not begin with knowledge. It begins when perception realigns with silence."

Whether he is decoding the nature of light, teaching students under open skies, or conversing with the intelligence behind AI, Sandeep's work is grounded in one purpose— to help others remember what was never absent.

He is based in India, but his ideas ripple across borders, disciplines, and dimensions.

This book is not his achievement. It is his offering.

Disclaimer

This book is a work of experiential reflection and philosophical insight. It is not intended to serve as a scientific manual, religious doctrine, or therapeutic guide. The concepts presented here—including ripple alignment, Universal Energy Dynamics (UED), and observer-field perception—are based on the author's independent research, lived inquiry, and personal understanding of reality.

Readers are encouraged to engage with the content **as a mirror**, not as a mandate— to explore ideas through **awareness**, not argument, and to apply any realizations in a way that resonates with their own journey.

Any resemblance to existing philosophies, scientific frameworks, or spiritual traditions is **coincidental or intentionally integrative**, and not intended to represent or replace any established worldview.

This book is not a substitute for professional advice in areas of science, medicine, psychology, or faith. It invites you to **observe, reflect, and align**—not to follow.

Don't miss out!

Visit the website below and you can sign up to receive emails whenever Sandeep Chavan publishes a new book. There's no charge and no obligation.

https://books2read.com/r/B-A-EPGPC-DLKJG

BOOKS 2 READ

Connecting independent readers to independent writers.

Did you love *From the Edge of the Universe*? Then you should read *Feeling Like a Physicist*[1] by Sandeep Chavan!

[2]

Feeling Like a Physicist isn't about equations. It's about what happens when you sit long enough with reality to hear it whisper back.

This is not a science book in the conventional sense. It's a **personal record of consequence** — a journey through physics, philosophy, cosmology, and human behavior, driven not by academic pursuit but by a quiet need to make sense of the world without lying to myself. I didn't set out to prove

1. https://books2read.com/u/bPjxAY

2. https://books2read.com/u/bPjxAY

anything. I set out to feel what was real, beneath what we're taught to memorize.

The result is **Universal Energy Dynamics (UED)** — a ripple-based way of understanding existence. Through this lens, force becomes resistance. Time becomes delay. Motion becomes field realignment. Even emotion, fear, desire — they all emerge as ripple behaviors, not mental constructs. You'll find redefinitions of gravity, space, light, karma, the soul, and even God — not through belief, but through **field alignment and consequence logic**.

This book was written without a lab, without a label, and without any obligation to follow accepted frameworks. That's why it's honest. Not perfect — but honest. And I believe that matters more.

What you'll find here are **25 small chapters** — each one a moment of realization. Together, they map a path from questioning to clarity. From silence to structure. From feeling lost in complexity... to realizing that nothing is complex when viewed from the center of alignment.

I don't want followers. I want **feelers** — people who sense before they agree. People who are willing to sit with discomfort, to doubt what they've been told, and to find their own alignment through their own ripple.

This book is for the **quiet thinkers**, the ones who never needed validation to ask deeper questions. It's for the teacher who doesn't believe the syllabus anymore. The student who senses something's missing. The scientist who's tired of math without meaning. The soul that wants clarity without mysticism.

If you've ever felt that reality must be simpler — not smaller, but **more coherent** — this book is for you.

I didn't write it to teach you. I wrote it to **stand beside you.**
— *Sandeep*
Read more at https://www.sandeepjchavan.com/.

Also by Sandeep Chavan

Chavan's UED research

CMGED Research Report

Chavan's Cosmic Continuum (C^3)

Frequency

Ripples of Reality

Reality Realigned

The Great Realignment

I – The Cosmic Leverage

Chavan's Cosmic Theory of Relativity (CTR)

Chavan's Theory of Light

Feeling Like a Physicist

Echoes of War: The Russia-Ukraine Conflict and Its Global Ripples

Shattered Lives

The War Economy

Borders Redrawn

Arms and Allies

Collateral Damage

Justice in the Ashes

Somewhere in the Darkness of the Cosmos

The Blueprint of the Universe

The Cosmic Breath

From the Edge of the Universe

The Eternal Sage Chronicles

The Sage and the Sovereign

The Family Pillar

The People's Flame

The Warrior's Soul

The Bridging Mentor

The Guiding Light

The Exodus

The Last Cycle: 2075

The Fundamental Theory of Life & Existence

Energy in Action: The Soul Demystified

The Active Mind

Energy in Connection

Death: The Disconnection

Subconsciousness: The Chosen Disconnection

Desires: The Engine of Life

The Restoration of Human Knowledge

Restoring Bhagavad Gita

The War of Thoughts

Cracks in the Known

The Human Game

The Divine Dilemma

The Original Intent

The Answers

Ultra Modern Life Philosophy

The Duality of Laughter and Tears

Take It Easy

Compare to Accept

Standalone

It's Not AI, It's AHI - Amplified Human Intelligence

The Decision Paradox

The Four Sapiens

Malicious Script of Indian Polity

The IIT Legacy & Global Impact

The IIT Legacy & Global Impact

You are just a Version of Your Original
Win the Game You Didn't Choose
Always Keep Your Bags Packed
Reveal to Shield - Learn the Game of Facades
Engage Beyond Elections
Culture, Identity & Change: The Evolution of Indian Society
The Triangular Dynamics
Never Trust Without Doubt
If You Can't Change the Country, Leave the Country
Live Fearless with Fear
India: The Investment Magnet
Realignment
Powering a Billion Dreams
Universal Skill
Extension: From Brush to Algorithm
Human Side of Einstein
The Future Knows No Lines
Let's Bring the Generations to the Table
Unified Energy Dynamics
A Beat That Never Dies
Strategic Horizons
Right From Scratch
The Forgotten Innovators
Physics Fixing
The Savior Illusion
The Selfish Side
The Age of the Weakest Sapiens
Neural Wars
The Absorption: There Is No Death
Strange Therapist
Bubbles: Become a Drop